MW01702025

Doctor Kaboom's
Bonkers Book of Gross Facts

Bonkers Facts!
Volume 1

Hi! I'm Izzy Sparks, engineer-in-training, future astronaut, world-class tinkerer, and proud niece of the one and only Dr. Ignatius Kaboom. Yes, that Dr. Kaboom. And no, I am not responsible for the time he melted half a refrigerator. I only encouraged him a little. Okay, maybe a medium amount. Look, good science requires enthusiasm, okay?

I love inventing things, fixing things, and figuring out why things work (or why they suddenly don't). When I'm not helping my uncle test slime samples or recalibrate his "extremely safe-ish" gadgets, I'm building my own machines, like my time-stabilized wrench or my anti-sock-losing device (still in beta). One day, I'm going to space to explore the entire universe. Until then, I'll explore the grossest, weirdest, slimiest corners of Earth right alongside you.

Let's go learn something

AWESOME!

If Izzy says "I have an idea," immediately locate the fire extinguisher.

No ticky, no touchy.

Doctor Kaboom's Safety Rules

Look, don't touch
Wild animals bite!

Never touch a wounded animal!

Give bugs space
They bite too!

Find a tick? Tell an adult.
No yanking, no tweezers, no heroics.

Science experiments mean
adult supervision.
Always. Even the easy ones.
Especially the fun ones.

This book is for learning.
Don't do anything without a grown up

Curiosity is awesome.
Just keep it safe.

Izz

Hey, Uncle Iggy, did you leave the spider cage open again? I can't find the giant tarantula.
– Iz

HOW TO USE THIS TOTALLY BONKERS BOOK

1. Look for the BIG FACT!

Every spread has *one giant, awesome, slightly disgusting truth* in huge letters. *That's the main idea. Your brain will love it.*

2. The smaller facts?

Those are the bonus gross-outs!
Read them in any order. Left to right, right to left, upside down.

3. Character Reactions = Hints!

Izzy's notes show what *kids* might wonder.
Dr. Kaboom's comments show what *grown-up scientists* think. Usually. Sometimes he's screaming.

4. Look for diagrams, zoom-ins, and tiny arrows.

They show how things actually work inside the animal world.
Real science → real awesome → real yuck.

Dr. Kaboom reads them alphabetically.
He's... like that.

5. You can read this book in ANY order.

Start with bugs, history, barf, or poop. Nature doesn't organize itself, so neither did we!

6. Safety First!

Gross facts are for reading, not re-enacting. If you're tempted to test anything...
→ Ask an adult.
→ Or ask Dr. Kaboom. He will say "NO."

No touching wildlife, bugs, or other gross things.
Definitely no frog-licking.
Experiments = grown-up required
Also: Don't try any animal behaviors or historical treatments. Ever. Learning ≠ doing.

7. Most important:

Have fun, stay curious, and enjoy learning the weirdest wonders of Earth!

Animals are magnificent marvels of evolution - graceful, noble, inspiring!

They also sneeze slime, shoot stink, drool waterfalls, eat horrifying things, and produce…and somehow ALWAYS wait until I'm standing directly in front of them. And they produce substances. SO. MANY. SUBSTANCES.

I adore studying animals, preferably from a polite and respectful distance. Unfortunately, animals rarely understand "polite and respectful distance." Some get curious. Some get offended. Some get very, VERY close. Which is why, if you'll excuse me…

WHY IS THAT GORILLA LOOKING AT ME

L I K E T H-

W H Y I S I T G E T T I N G

C L O S E R-

w h y i s i-

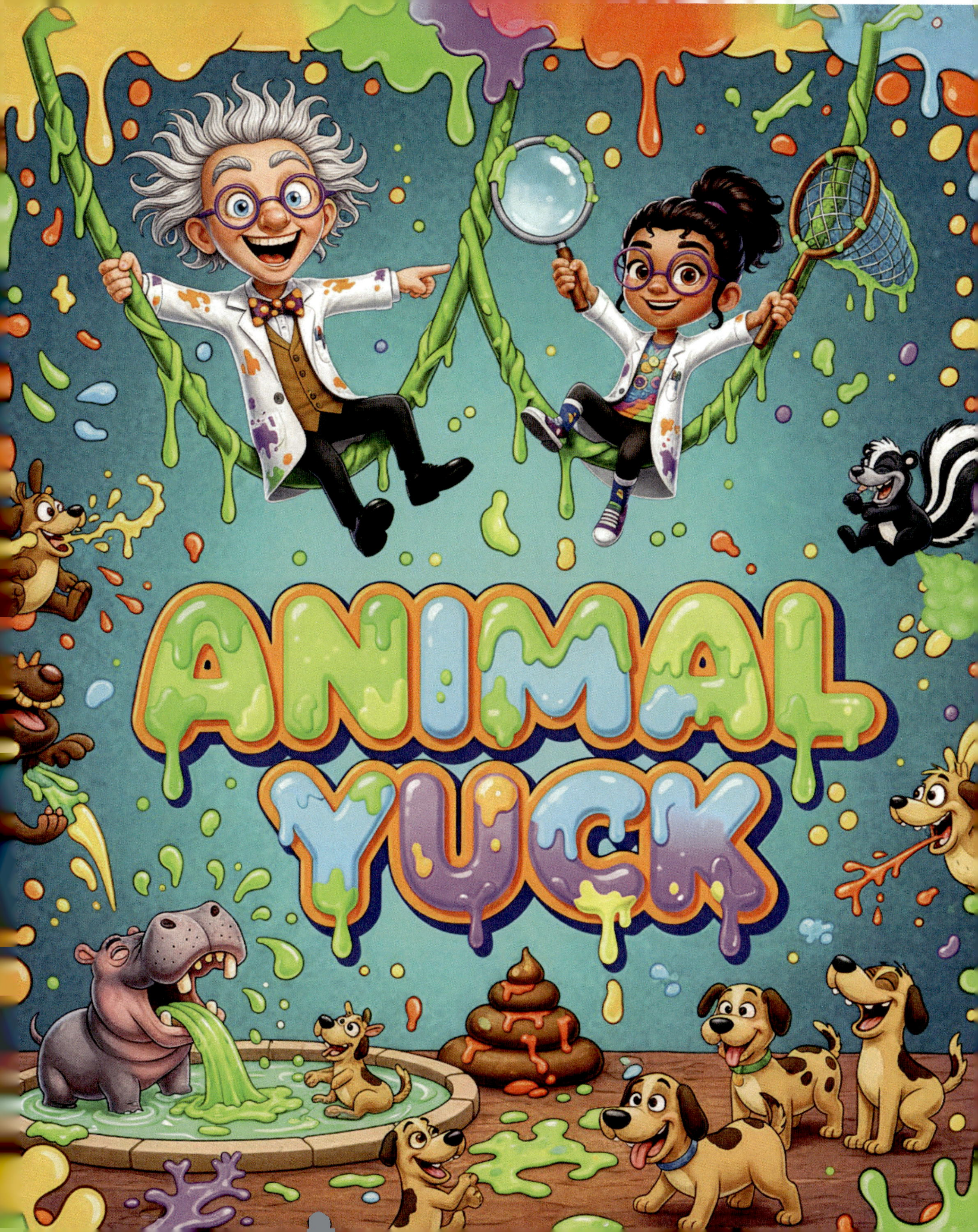
ANIMAL
YUCK

Big Fact:

Vultures poop on their own legs to stay cool!

Their droppings help cool them down in hot weather and kill bacteria from all that rotten meat.

Their stomach acid can dissolve metal. Nature's strongest disinfectant. Built different.
They eat bones too. Nothing goes to waste at the vulture buffet.
They're actually super clean birds. Weirdly enough—poop baths work!

Penguins shoot poop up to four feet away!

They blast it from their nests to keep things tidy—like high-pressure hoses in tuxedos.

The poop is pink from krill.
Even their waste is festive!

Scientists study it from space.
Poop piles show where colonies live. No cap.

Baby penguins beg for it.
Yep, parents feed them pre-digested fish slime.

HiPPOS MAKE POOP SHOWERS TO MARK TERRiTORY!

They spin their tails while pooping to fling it everywhere—talk about over-sharing.

They do it in water too.
River rave, hippo edition!
Their poop feeds fish.
Disgusting, but efficient.
They sweat pink sunscreen.
Stylish and slimy!

Bees vomit to make honey!

They collect nectar, store it in a honey stomach, and **regurgitate** it into the hive. Deliciously disgusting!

HONEY NEVER SPOILS.
It's bacteria-proof thanks to bee spit enzymes.
BEES FAN IT TO DRY IT OUT.
Tiny buzz-powered dehydrators!
IT TAKES 12 BEES' LIVES TO MAKE A TEASPOON.
Respect the puke, people.

Komodo dragons drool deadly spit!
Their saliva is full of bacteria and venom that helps them take down prey. One bite, slow doom.
That's one mouthy monster!

THEY CAN SMELL BLOOD FROM MILES AWAY.
REPTILE CSI!
THEY'RE THE WORLD'S LARGEST LIZARDS.
Ten feet of fang and funk.
THEY EAT 80% OF THEIR BODY WEIGHT AT ONCE.
Thanksgiving lizard edition.

Some frogs carry babies in their backs!

Darwin's frogs swallow eggs and grow tadpoles in their mouths until they hop out. Surprise!

OTHERS GROW BABIES UNDER SKIN.

Backpack frogs, but grosser.

FROG SLIME HELPS HEAL WOUNDS.

Slime with purpose!

THEY BREATHE THROUGH SKIN.

Breathing jackets!

Bats pee while flying!

Gravity never takes a break. They let it rain mid-flight to stay light and agile.

They hang upside down to rest.
The pee drains away—good planning!
Their poop is called guano.
Used as fertilizer and explosives! Bussin business.
Bats eat half their weight in bugs nightly.
Midnight pest control.

Elephants use snot showers to cool down!

They spray water and mucus from their trunks to keep cool—trunks double as **slime** cannons.

THEIR EARS ACT AS FANS
Built-in air conditioning
THEY CAN SMELL WATER FROM MILES AWAY.
Trunk GPS!
THEY FLAP EARS WHEN HAPPY.
Science's biggest wiggly bois.

SEA BIRDS DRINK OCEAN WATER AND SNEEZE OUT THE SALT!

Special glands near their noses remove salt and spray it out. Gesundheit, seagull!

THEY CAN DRINK SEAWATER SAFELY.
No dehydration here.
ALBATROSSES TRAVEL YEARS AT SEA
Sneeze-powered navigation!
SALT CRYSTALS FORM ON THEIR BEAKS.
Instant science glitter.

EXPERIMENT ZONE

STOP! Pay attention because I'm about to show you something that will BLOW YOUR MIND. Ready? You're about to make raisins DANCE. No music. No strings. Just SCIENCE! I once watched raisins dance for three hours straight. I forgot to eat lunch. I forgot my own middle name. Best day of my life.

KABOOOOOOOM!

MATERIALS:

- 1 tall clear glass or jar
- Clear carbonated soda (lemon-lime works great—you want to SEE what's happening)
- 6-8 raisins (the wrinklier, the better)

Adult supervision required.
Blame the Fun Police.

INSTRUCTIONS:

Step 1: Fill your glass about ¾ full with fresh, fizzy soda. It MUST be freshly opened. Flat soda is a dance floor with no music.

Step 2: Drop in your raisins. Watch them sink to the bottom like tiny wrinkled submarines.

Step 3: Wait 30-60 seconds. Keep watching.

Step 4: WITNESS THE DANCE. The raisins will rise to the surface, hang out for a moment like they're taking a breather, then sink back down. Up. Down. Up. Down. This will continue until the soda goes flat—sometimes 20-30 minutes of *non-stop raisin disco*. Welcome to the Raisin Rave. No glow sticks required.

Bonus: Try using pasta, dried cranberries, or popcorn kernels! Which one dances best?

EXPERIMENT ZONE

THE GROSS SCIENCE:

Here's what's actually happening in your glass of soda-powered magic:

Soda is full of **carbon dioxide (CO_2)**—that's what makes it fizzy. Those tiny bubbles are constantly forming and rising to escape into the air.

Science Terms:

Buoyancy: The upward force that makes things float
Density: How heavy something is compared to its size
Carbon dioxide: The gas that makes soda fizzy (and makes you burp)

Why raisins dance best:

- wrinkle folds = bubble pockets
- bubbles attach → raisin gets lighter
- bubbles pop → raisin sinks again

The number of bubbles that attach depends on surface area. That's why wrinkly raisins work better than smooth grapes. They have more places for bubbles to stick. Engineers use this same idea when designing things that need to trap air, like life jackets!

When the raisin hits the surface, the bubbles POP away. Without them, it sinks again, until new bubbles grab on and lift it back up.

This is the same principle that explains why YOU float in water when you take a deep breath (your lungs fill with air) and sink a little when you breathe out.

Okay, okay — DON'T freak out — but this chapter is full of things that want to slurp your blood like a science juice box. Cool, right?!

Uncle Kaboom made me wear ALL this gear: rubber gloves, a garlic necklace, and a helmet that goes BZZZT every three seconds. I think he just grabbed everything labeled "protective" and threw it at me.

But HERE'S THE THING: bloodsuckers are amazing. Tiny flying robots with needles for faces! Wings that beat a thousand times per second! They pollinate flowers, feed bats, and keep whole ecosystems running! Yeah, sometimes they turn your arm into itchy connect-the-dots. But STILL. Grab your bug spray and LET'S MEET THE SLURPIEST CREATURES ON EARTH!

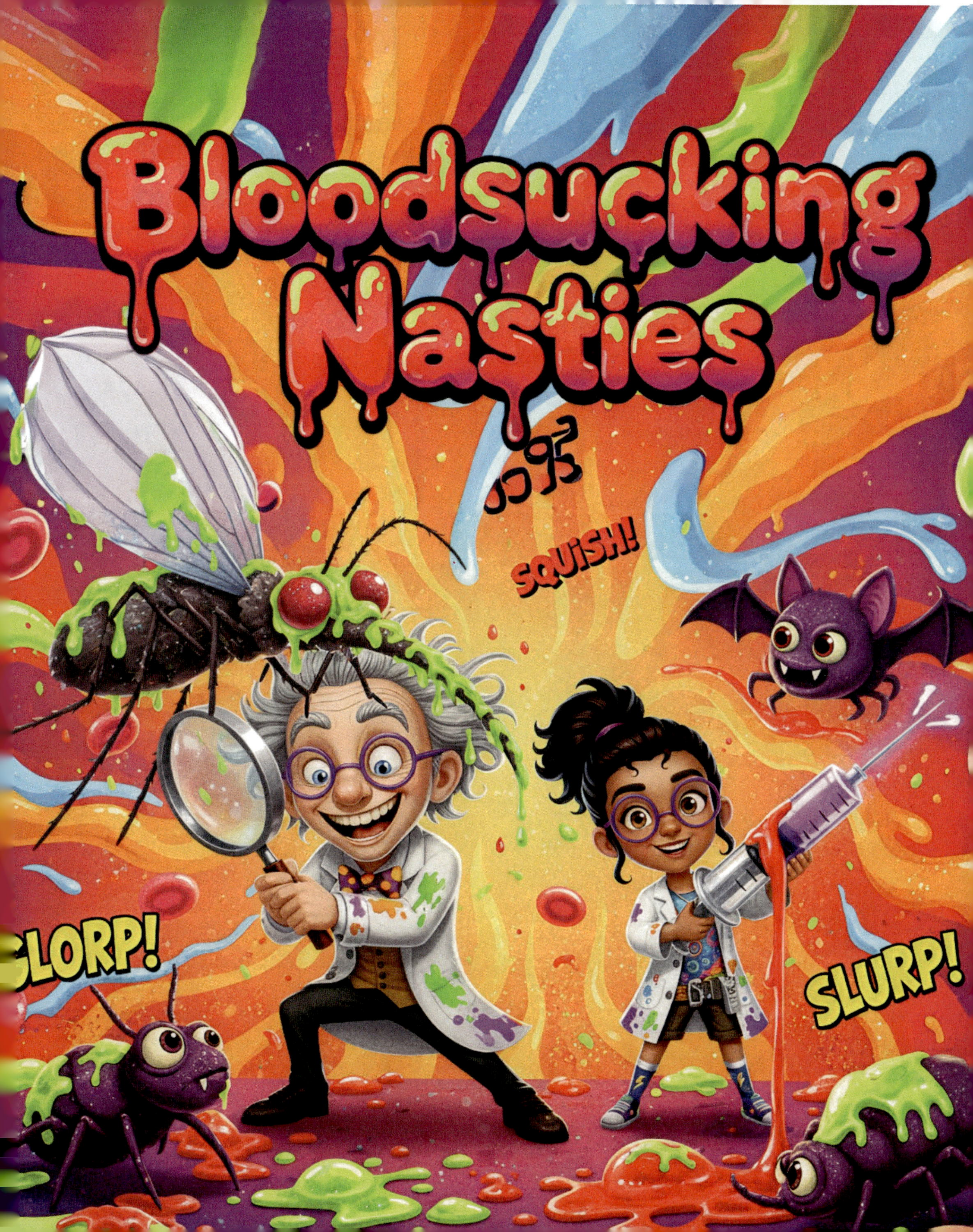
Bloodsucking
Nasties
SQUISH!
LORP!
SLURP!

Some butterflies sip blood, sweat, and even tears!

A few butterfly species drink from mud, wounds, sweat, and animal tears to get extra minerals they can't find in flower nectar.

SMALL FACT 1
THEY LOVE SALTY SNACKS.
Sweat gives them the minerals they crave.
SMALL FACT 2
TEAR-DRINKERS
They land on turtles and crocodiles to sip tears!
SMALL FACT 3
MUDPUDDLE PARTIES
They slurp nutrients from mud, pee, and poop.

Mosquitoes drink blood to grow eggs!
Only females bite—they slurp up blood proteins to fuel baby mosquito production.
Tiny, winged vampires of chaos!
BLOOD

They find you by smell.
Your stink guides them like GPS.
They spit first.
Saliva numbs you before the sip.
They prefer sweaty feet.
Foot-funk = mosquito magnet.

Leeches suck blood with three jaws!

Their bite leaves a perfect Y-shaped mark while their saliva keeps your blood flowing.

THEY NUMB THE BITE.
You don't feel them latch on.
THEY CAN SWELL HUGE.
Up to ten times their weight!
DOCTORS STILL USE THEM.
To restore blood flow in surgeries.

HEADLINE: Ticks drill into skin with barbed mouthparts!
EXPLANATION:
Their backward spikes lock them in place while they drink for days.
Do NOT boop the tick snoot!

They inject cement-like glue.
To stick even tighter.

They balloon with blood.
Engorged to grape-size.

They detect breath.
CO_2 tells them where dinner is.

VAMPIRE BATS DRINK BLOOD LIKE JUICE BOXES!

They make a tiny cut, lick the blood, and use heat sensors to find the best spot.

THEY SHARE MEALS.
Generous little goths.
THEY WALK ON THEIR THUMBS.
Crawling creepers.
THEIR SPIT KEEPS BLOOD RUNNY.
Nature's anti-coagulant.

Fleas can drain 15 times their body weight!

These tiny jumpers pierce skin with needle mouths and feast nonstop.

They jump like superheroes.
Up to 200x their body length!
They lay hundreds of eggs.
Instant infestation mode.
They spread diseases.
Including plague—yikes.

Some parasitic wasps drink the blood of their hosts!

They sting, stun, and then slurp hemolymph—the insect version of blood.

They zombify victims.
Mind-control realness.
Larvae eat hosts alive.
Terrifying daycare service.
Some protect gardens.
Nature's brutal pest control.

Ticks aren't the only bloodsuckers—mites join the party!

Some mites feed on birds, reptiles, and even mammals, sneaking between feathers and scales.

THEY'RE MICROSCOPIC.

Tiny vampires unseen.

THEY SWARM FAST.

A thousand bites before sunrise.

BIRDS DUST-BATHE.

Nature's mite repellent.

Oxpecker birds drink blood while cleaning rhinos!

These birds hitch a ride on rhinos, zebras, and buffalo, eating ticks and sipping the blood left behind from bite wounds.

They're pest control pros. They pick off ticks, fleas, and bugs all day.

They warn of danger. Their hiss-alarm tells big animals trouble is near.

They keep wounds open. More blood snacks for them—less comfy for rhinos.

Here's a question that's going to change how you see the world: What IS snot, actually? It's not just gross goo your nose makes to annoy you. It's a DEFENSE SYSTEM. Your nose produces about a liter of mucus EVERY DAY-most of which you swallow without knowing it. You're welcome for that information. Now let's make SNOT!

KABOOOOOOOM!

MATERIALS:

- 1 cup warm water
- 1 packet unflavored gelatin (Knox brand works great)
- 2-3 tablespoons light corn syrup
- Green and yellow food coloring
- Mixing bowl
- Fork or whisk

Adult supervision required.
Or the Safety Goblins will get you.

INSTRUCTIONS:

Step 1: Heat 1 cup of water until warm (not boiling. We're making snot, not soup).

Step 2: Sprinkle the gelatin packet over the warm water. Let it sit for 2 minutes to "bloom." It'll get weird and clumpy. This is correct.

Step 3: Stir with a fork until mostly dissolved. Some lumps are fine. Snot isn't smooth anyway.

Step 4: Add the corn syrup. Stir until combined. Watch the mixture get THICK and STRETCHY.

Step 5: Add 2-3 drops of green food coloring and 1 drop of yellow. Stir until you achieve "sick day" color.

Step 6: Let it cool for 5-10 minutes. The longer it sits, the thicker it gets.

Step 7: Pick it up. Stretch it. Let it drip through your fingers. Congratulations: you've made boogers!

EXPERIMENT ZONE

THE GROSS SCIENCE:

Your nose isn't just a face hole. I’s a highly sophisticated air filtration system, and MUCUS is the star of the show.

Mucus is made of:

- **Water** (95%)
- **Glycoproteins** called MUCINS (the sticky, stretchy part)
- **Antibodies** (germ fighters)
- **Enzymes** (germ killers)
- **Salt** (which is why boogers taste salty. Don’t pretend you don't know!)

The **gelatin** in our recipe mimics **mucins**. They’re both **polymers**, which means they're made of long chain-like molecules that stretch and flow. That's why real snot (and our fake version) can be both liquid AND solid-ish at the same time.

Polymers are EVERYWHERE—in rubber bands, in plastic, in the slime we just made, and in your actual boogers. They're molecules that link together in long chains. The longer the chains, the stretchier the stuff. That's why our corn syrup helps—it lets the gelatin chains slide past each other more easily.

The **corn syrup** increases viscosity, a fancy word for "thickness." It makes our slime flow slowly, just like the real thing.

Real mucus traps dust, pollen, bacteria, and viruses before they can reach your lungs. The tiny hairs in your nose (cilia) then sweep all that trapped gunk toward your throat so you can swallow it. Yes, you eat your own boogers constantly. Your body doesn't care about table manners.

Why is snot sometimes GREEN? When you're sick, your immune system sends white blood cells to fight infection. Dead white blood cells contain an enzyme with iron in it, which turns mucus green. Green snot = your body won.

Excellent! You've arrived at my favorite section of all: BODY GUNK! Nature's sticky, slimy, slippery way of keeping everything running smoothly… or at least hilariously. Every creature on Earth is basically a walking chemistry lab full of fluids that squirt, drip, ooze, bubble, or erupt at the worst possible moment. Mucus, drool, stomach slime, ear wax, eyeball goo. It's all part of the marvelous, slightly squishy miracle of biology! Think of it as nature's maintenance slime. Disgusting? Yes. Useful? Absolutely. Sticky? ALWAYS.

Now, I must warn you that learning about bodily gunk is not for the faint of heart or the clean of shirt. Scientists like myself face dangerous levels of splatter on a daily basis. Do you know how many lab coats I've ruined? Fourteen. In one week. But fear not! With a strong stomach, a curious mind, and perhaps a poncho, you'll soon discover that grossness is just science wearing a silly costume. Ready? Deep breath. (Preferably through your mouth.) Let us journey into the delightfully icky world of BODY GUNK!

KABOOOOOOOM!

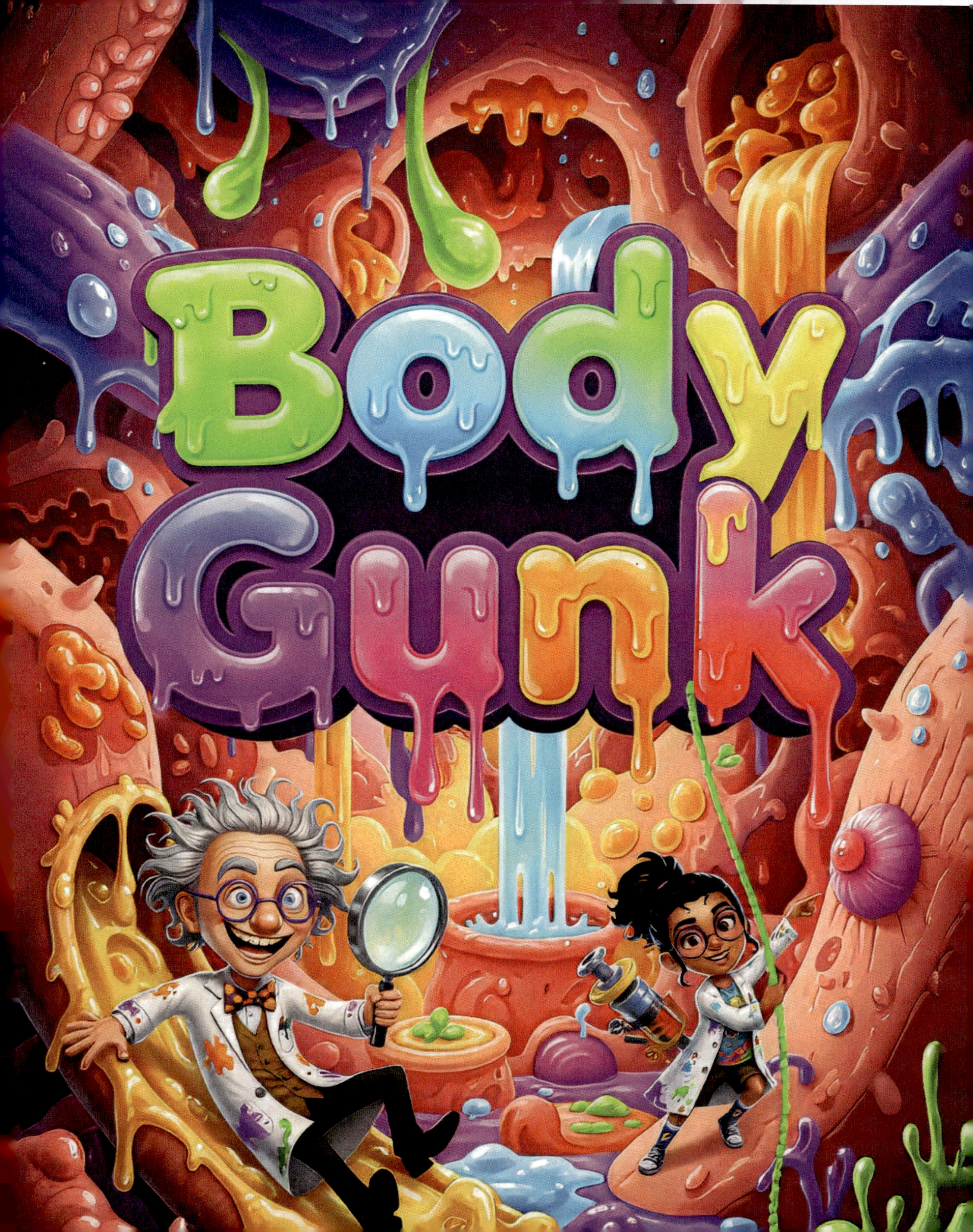
Body
Gunk

Your body makes a whole liter of snot every day!

That goo traps dust and germs before they get to your lungs. It's one of your body's defense systems!

Boogers are just dried-up snot catching dust.

Your nose basically makes sticky flypaper to keep out bad stuff. Ew-genius!

Sneezes sneezes blast at 100 miles per hour!

That's faster than a race car. Your nose is a turbo-charged cannon!

Your nose hairs are nose ninjas.

They block dirt, pollen, and tiny bugs like fuzzy little bouncers.

Earwax is your ears' homemade shield!

That waxy goo keeps dirt, bugs, and germs out. Think of it as ear armor!

EARWAX CLEANS ITSELF OUT NATURALLY.
No cotton swabs needed! Your jaw moves it out when you talk or chew!
THERE ARE TWO TYPES OF EARWAX: WET AND DRY.
It depends on your genes! It's the weirdest family trait ever.
Smelly
EARWAX CAN SMELL DIFFERENT DEPENDING ON YOUR DIET.
Eat more onions, smell more... interesting.

Your belly growls because it's talking to itself!

When your stomach is empty, it squeezes and gurgles to move leftover food around. It's like belly gossip!

Your stomach is strong enough to melt TEETH!
Do not eat your teeth!
Your tummy replaces its lining every few days.
It's constantly remodeling—your stomach is its own construction site.
That rumble before lunch is called borborygmi.
It's a fancy word for "hungry thunder." Say it three times fast!
RUMBLE!
GRRRR!

Your skin flakes off about a million times a day!

You're constantly making new skin cells, so the old ones dust off. You're basically a walking snowstorm!

Most household dust is old skin cells.
So yeah—your room's kinda made of... you.
You grow new new skin every 27 days.
That's like getting a free upgrade every month!
Your skin is your biggest organ.
It covers about 20 square feet. You're wrapped in science!

Your feet are Sweat factories!

Each foot has over 250,000 sweat glands. That's a quarter million tiny sprinklers making foot soup!

Feet can sweat half a pint a day,

That's enough to fill a small soda can—ew, don't drink it!

Sweat itself doesn't smell.

It's bacteria partying in your sweat that makes the stink!

Some people's feet glow under UV light.

Certain bacteria on your skin actually glow. Disco toes!

FARTS ARE TiNY GAS EXPLOSiONS iN YOUR GUTS!

When bacteria in your intestines eat food, they make gas. It's science's stinkiest side effect!

FART FACTS!

Most farts are just air you swallowed.

So technically, you're burping backward!

PooF!

Farts are mostly odorless.

Only about 1% of the gas actually smells. But oh, that 1% works hard.

You fart around 14 times a day.

That's normal. Congratulations, you're alive!

KABOOM!

Vomit is your body's emergency eject button!

If you eat something bad, your body hits the eject lever to blast it out fast. Better out than in!

YOU MAKE EXTRA SALIVA BEFORE YOU PUKE.

It protects your teeth from stomach acid—like a gross mouth helmet.

SOME ANIMALS EAT THEIR OWN VOMIT.

Yeah. Science doesn't always make sense.

VOMIT CAN SHOOT SEVERAL FEET.

Your stomach muscles are *that* powerful. Ewlympic-level strength!

YOUR SPIT HELPS YOU TASTE AND DIGEST FOOD!

Saliva breaks down food before it even hits your stomach. It's your mouth's secret superpower!

You make about a quart of spit every day.
That's enough to fill a water bottle. Hydration... sorta.
Spit helps heal wounds.
It's got special proteins that help cuts close faster.
Some animals spit to defend themselves.
Camels and llamas are the spit-slinging champions of the animal world!

Sweat keeps you from overheating!

When your body gets hot, sweat cools you off as it evaporates. You're a self-moisturizing machine!

YOUR HANDS AND FEET HAVE THE MOST SWEAT GLANDS.

That's why your palms get clammy before a big test!

HIPPOS SWEAT PINK!

Their sweat acts as sunscreen and bug repellent. Fancy!

DANGER

SOME PEOPLE BARELY SWEAT AT ALL.

It's a rare condition called hypohidrosis. Sounds cool, but it's actually dangerous!

EXPERIMENT ZONE

Raise your hand if you've ever picked a scab. Now put your hand down because that's gross and also because I can't see you. The point is: scabs are DISGUSTING and AMAZING. They're your body's emergency bandage, built right there on your skin using nothing but blood and chemistry. Today we're going to make our OWN SCABS! No actual bleeding required. You're welcome.

KABOOOOOOOM!

MATERIALS:

- 1 tablespoon petroleum jelly (like Vaseline)
- 1 tablespoon all-purpose flour (CRITICAL—do not skip this)
- Cocoa powder (unsweetened works best)
- Instant coffee grounds or finely crushed coffee
- Red gel food coloring
- Small mixing container
- Toothpick or small spatula for application

Adult supervision required.
The Department of Worried Adults required this.

INSTRUCTIONS:

Step 1: Combine petroleum jelly and flour in equal parts (1 tablespoon each). Mix until it forms a dough-like putty. This is your "skin putty" base—without the flour, you just have a greasy mess.

Step 2: Add cocoa powder a little at a time until you reach a brownish-red color. Start with about ½ teaspoon. Real scabs are darker than you think.

Step 3: Mix in a pinch of coffee grounds for texture. Real scabs have that rough, crusty surface—the coffee grounds nail it.

Step 4: Take a small blob and press it onto your skin where you want your "wound." Shape it with your fingers—think about how real scabs form in irregular shapes.

Step 5: Use a toothpick to dab red gel food coloring around the edges (the "fresh" part where it's still healing).

Step 6: Let it set for 5 minutes. Admire your disgusting handiwork. Terrify a family member.

EXPERIMENT ZONE

THE GROSS SCIENCE:

When you get a cut, your body launches an INCREDIBLE emergency response:

STEP 1: THE ALARM Damaged blood vessels immediately start leaking. Your body goes "OH NO" and releases chemical signals.

STEP 2: PLATELET PARTY Platelets (tiny cell fragments in your blood) rush to the scene. They're sticky little guys who clump together to form a "platelet plug"—like putting your finger over a hole in a water balloon.

STEP 3: THE FIBRIN NET Your blood contains a protein called fibrinogen. When injury happens, it transforms into FIBRIN—long, sticky protein strands that weave together like a net. This net traps platelets and red blood cells, forming a solid clot.

STEP 4: SCAB TIME The clot dries out and hardens. That's your scab, a temporary shield made of dried blood cells, fibrin, and platelets.

STEP 5: HEALING UNDERNEATH While the scab protects the surface, new skin cells are growing beneath it. When they're done, the scab falls off. (Or you pick it and start the whole process over. STOP PICKING.)

We're using flour + petroleum jelly to create what special effects artists call 'skin putty.' The flour gives it structure so it holds its shape, while the petroleum jelly keeps it sticky and moldable. It's the same basic technique used in movies. Just way cheaper!

Why is dried blood brown? Fresh blood is red because of iron-rich hemoglobin. When blood dries, the hemoglobin oxidizes (reacts with oxygen) and turns brown—just like rust. Because it's basically the same chemistry!

People today may be pretty gross, but OH BOY — humans throughout history were operating on a whole different setting. Welcome to GROSS HISTORY, where ancient stinkiness meets medieval nastiness and early-scientist "experiments" that probably should've been illegal. Spoiler: the past was REALLY dirty. Like... unbelievably dirty. Like "did you wash your hands this month?" dirty.

But that's what makes it awesome to explore! From strange medical tools to questionable hygiene habits to some of the most bizarre inventions ever created, we're diving into humanity's ickiest hits. Grab your time-travel goggles and hold your nose — because this trip through history is about to get hilariously messy.

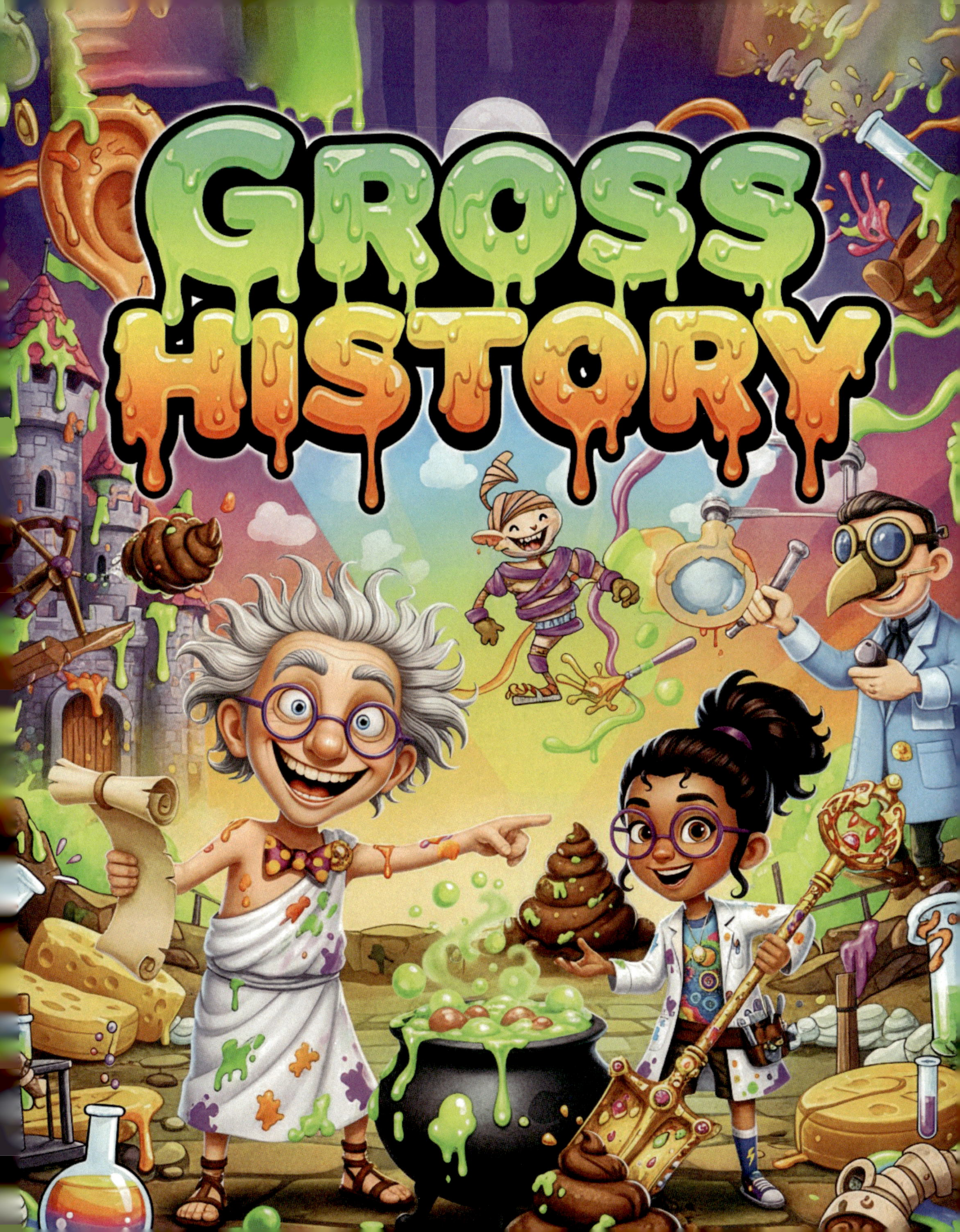
GROSS
HISTORY

ANCIENT ROMANS USED PEE TO WASH CLOTHES!

They collected urine because it contains ammonia—a natural cleaner. Laundry day never smelled so bad!

Urine was taxed by the emperor.
Yes, there was literally a pee tax. Lowkey wild.
Roman laundromats were called fullonicae.
Workers stomped clothes in tubs of warm pee. Ew!
They also used pee to whiten teeth.
Minty fresh? Not so much.

Doctors used to taste urine to diagnose illness!

Before lab tests, they'd sample pee to check sugar levels. That's how they found diabetes—sweet pee!

It was called uroscopy.

They even had pee color charts!

Sweet pee meant diabetes.

Science was rough before test strips.

Some wore pee-proof gloves.

Smartest doctors in the room.

People used to think bad smells caused disease!

Before germ theory, they believed "miasma"—evil air—spread sickness. So they sniffed flowers for protection.

Plague doctors wore bird masks.
The beak was stuffed with herbs to block smells.
Perfume was considered medicine.
Smelling good = staying alive! No cap.
They burned spices in homes.
Basically ancient Febreze.

IN THE MIDDLE AGES, PEOPLE RARELY BATHED!

Baths were thought to spread disease. Some folks bragged about not washing for years.

THEY COVERED SMELLS WITH PERFUME.
The first deodorant, sort of.
ROYALTY BATHED MORE THAN PEASANTS.
Cleanliness was a luxury.
THEY USED CLOTHS CALLED "DRY BATHS."
Wipe downs instead of water!

Toothpaste used to be made of ashes and crushed bones!

Before minty gels, people scrubbed with burnt shells, charcoal, and ground animal parts.

Ancient Egyptians made the first toothpaste.

With ingredients like ox hooves and myrrh.

Romans added mouse brains.

Mouth feel: horrible.

Charcoal toothpaste is making a comeback!

What's old (and gross) is new again.

Early surgeons didn't use anesthesia!

Patients were awake for operations. Ouch! Some surgeons were faster than race car pit crews.

They gave patients alcohol to dull pain.

Liquid courage in a bottle.

Amputations were timed events.

Speed was everything!

Anesthesia wasn't invented until the 1800s.

Science finally said, 'maybe don't scream?"

People used to eat mummies as medicine!

They ground up Egyptian mummies, thinking it would cure disease. Spoiler: it didn't.

Mummy powder was sold in apothecaries.
Ancient body dust for health!
It was mixed in drinks.
Smoothie of nightmares.
APOTHECARY JAR
Fake mummies were made when supply ran out.
Counterfeit corpses—yuck!

In the 1700s, wigs were full of bugs!

Powdered wigs attracted lice, fleas, and sometimes mice. **Scratch that!**

People scratched with fancy sticks.
Elegant itch relief!
Perfume hid the stench.
Everyone smelled like rotten flowers.
Some wigs caught fire from candles.
Fashion statement... and fire hazard.

Victorian doctors used leeches for everything!

They thought sucking blood cured fevers, headaches, and bad moods. The leeches made bank.

Doctors kept jars of leeches on desks.
Gross office décor.
Some people wore them as therapy.
Spa day, 1800s-style.
They're still used today—in surgery!
Modern science meets old-school slime.

Welcome to the wriggling, squirming, twitching world of CREEPY CRAWLERS! These tiny critters may look small, but trust me, they pack a truly heroic amount of grossness. Some crawl, some ooze, some have more legs than seems morally acceptable. I've been studying them my whole life - mostly because they keep climbing into my lab coat pockets when I'm not paying attention.

But don't worry! These remarkable little creatures aren't here to scare you (well, maybe a little). They're here to show you how utterly AMAZING the miniature world can be. So grab your magnifying glass and your bravery. Things are about to get creepy… and crawly… and spectacularly slimy!

KABOOOOOOOOM!

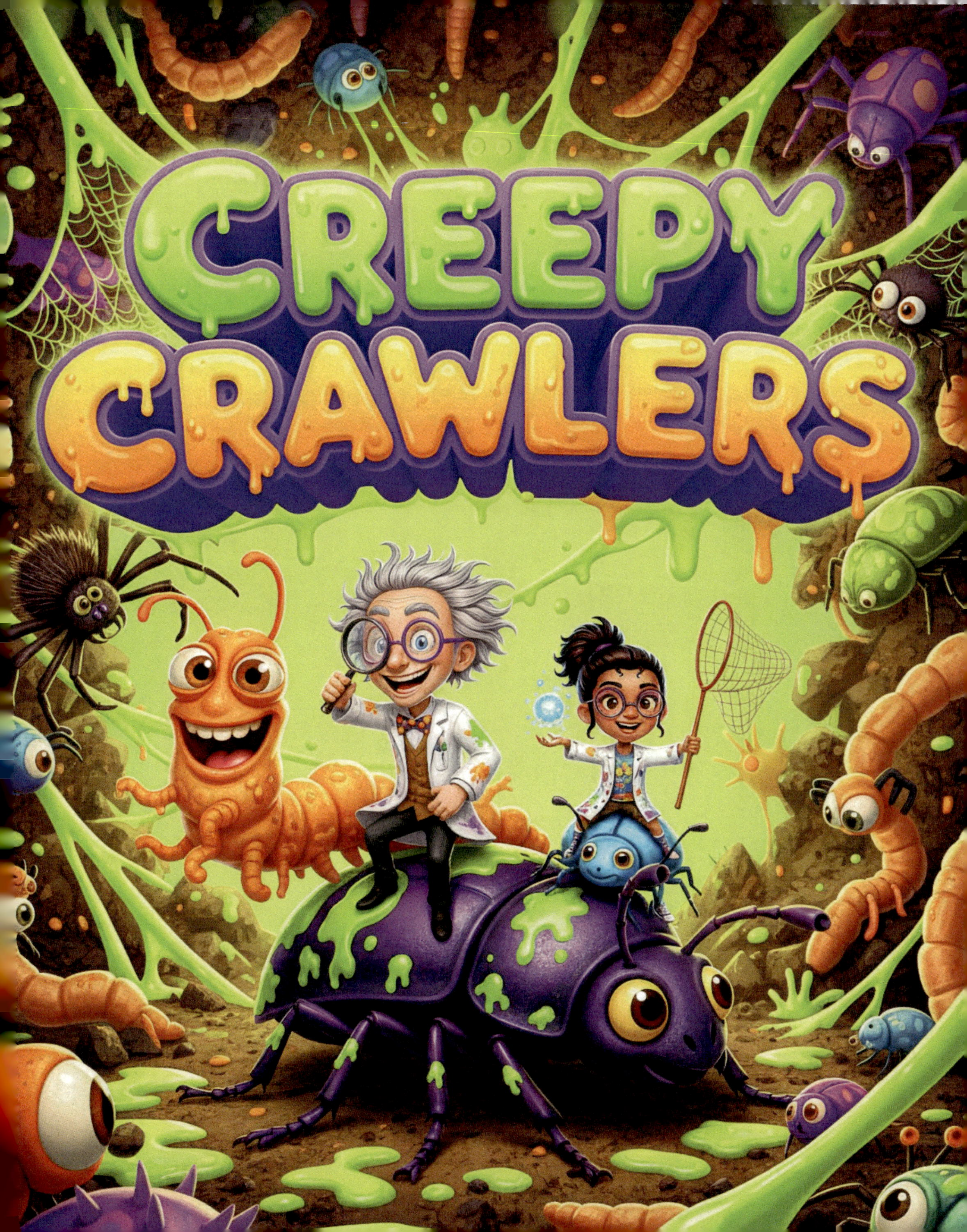
CREEPY
CRAWLERS

COCKROACHES CAN LIVE FOR WEEKS WITHOUT THEIR HEADS!

Their brains aren't in one place like ours—they've got nerves all over. They only die because they can't drink water!

Roaches breathe through tiny holes in their sides.
So even decapitation is just a mild inconvenience. No cap.
They've been around since dinosaurs.
Survival level: absolutely bussin.
They can run up to 3 miles per hour.
Basically, mini sprinters with bad manners.

Earthworms have five hearts!

They need them to pump blood through
their long, stretchy bodies.
Talk about overachievers!

Worms can regrow parts of themselves.
Cut one in half and—ta-- da!—you might get a new worm. Lowkey magic.
They eat dirt to get nutrients.
A truly down-to-earth diet.
Worms breathe through their skin.
Slimy skin = breathing machine.

Leeches were once used as medicine!

Doctors used them to suck out "bad blood." Gross but oddly effective in some cases!

Leeches have 32 brains.

And yet, none of them think this is sus.

Their saliva stops blood from clotting.

Nature's original anti-coagulant.

They can suck blood up to ten times their body weight.

Science's hungriest interns.

Maggots can clean wounds!

They eat only the dead tissue and leave the healthy stuff alone—little zombie doctors!

FLIES LAY UP TO 500 EGGS AT A TIME.

Each one could become a maggot medic. Squad goals.

MAGGOTS GROW SUPER FAST.

They can double in size in a day.

THEY BREATHE THROUGH TINY TUBES ON THEIR BUTTS.

Talk about multitasking!

Spiders eat their own webs to recycle silk!
It's how they get protein back to make new webs—zero waste living!
Web you believe it?!

Spider silk is stronger than steel.
Weight for weight, it's one of nature's strongest materials.
Slaps hard.
Some spiders can glide on the wind.
They spin silk parachutes to travel miles!
Male spiders bring snacks on dates.
It's romantic... unless she eats him.

ANTS NEVER SLEEP!

They take micro-naps instead—just a few seconds at a time. Always hustling!

Ants can lift 50 times their weight.

That's like you bench-pressing a car. Built different.

Ants have two stomachs.

One for eating, one for sharing. Ew, community stomach.

They farm other bugs for food.

Tiny farmers with questionable ethics.

CENTIPEDES HAVE VENOMOUS FANGS!

They inject venom into prey with their first pair of legs. Don't worry—they're more scary than deadly!

Doctor Kaboom's

BONKERS BOOK OF GROSS FACTS

By Dr. Ignatius Kaboom

Created & Illustrated by
August Trometer

A

BONKERS PRESS™

Publication

Doctor Kaboom's Bonkers Book of Gross Facts!

Created and illustrated by August Trometer
Written by Dr. Ignatius Kaboom

Bonkers Press™
Los Angeles, California
www.bonkerspress.com

All facts presented herein have been checked, double checked, re-checked, and validated. That said, if any slipped by us, we're very sorry.

ISBN: 979-8-9911675-4-3
Printed in the United States of America
First Edition

10 9 8 7 6 5 4 3 2 1

For Bond and Lyric,

my brightest sparks,
my favorite explorers,
and the future scientists who
will change the world.

Hello, my explosively curious friends! I am Dr. Ignatius Kaboom, the world's leading expert in weird, wild, and wonderfully disgusting science. If it oozes, slimes, splats, squirms, burps, bubbles, or smells so bad it makes your eyebrows curl, I've probably poked it with a stick. My life's work is discovering the hidden wonders of science and explaining them before they crawl away. Or explode. Or both. Usually both.

Some people say I'm a little overexcited. NONSENSE! Science is supposed to be exciting! My hair didn't get like this from sitting calmly in a chair. No, it got this way during an experiment involving slime, static electricity, fermented cabbage gas, and a very angry raccoon. Long story, but I survived! And so will you… probably.

So grab your goggles, tighten your bow tie, and let's explore the wonderfully icky universe of GROSS FACTS!

KABOOOOOOOM!

CENTIPEDES AREN'T INSECTS.

They're arthropods—fancy word for "creepy leg gang."

SOME GLOW IN THE DARK.

Because why not be terrifying *and* stylish?

THEY CAN REGROW LOST LEGS.

Free replacements, courtesy of evolution.

Beetles make up 40% of all insects on Earth!

There are over 350,000 species. If it crawls, crunches, or clicks, it's probably a beetle.

Dung beetles roll poop balls bigger than themselves.

Strong, smelly, and strangely proud.

Fireflies are beetles too.

They light up with chemical magic called bioluminescence. Glow-up goals.

Beetles have armor-like shells.

Nature's tanks, ready for bug battle.

Mosquitoes love stinky feet!

They're attracted to sweat smells, carbon dioxide, and warm bodies. You're basically a buffet!

Only female mosquitoes bite.

They need blood to grow eggs. The males just sip nectar.

They flap their wings 600 times a second.

That whine in your ear? Their love song.

They can detect you from 100 feet away.

You can run, but you can't hide—especially if you stink!

GROSS-ARY: A GROSS GLOSSARY

(And words your friends don't know!)

Bioluminescence (bye-oh-loo-mih-NESS-ence)
When a living creature makes its own light - like glowing bugs, jellyfish, or fish with spooky headlights.

Mucus (MYOO-kuss)
The sticky goop your body makes to trap germs and keep things moist.

Enzymes (EN-zimes)
Tiny protein helpers that speed up chemical reactions - like breaking down food in your stomach.

Hydrochloric Acid (hy-droh-KLOR-ik ASS-id)
Your stomach's super-strong acid that helps dissolve food (and could ruin a metal coin's day).

Bile (BYE-ul)
A yellow-green digestive juice from your gallbladder that helps break down fats.

Parasite (PAIR-uh-site)
An organism that lives on or inside another organism and steals its food or nutrients.

Proboscis (proh-BAH-sis)
A long, tube-shaped mouth used for sipping, slurping, or stabbing (looking at you, mosquitoes).

Buoyancy (BOY-un-see)
The upward force that makes things float - including dancing raisins.

Density (DEN-suh-tee)
How heavy something is for its size - a rock has more density than a marshmallow.

Carbon Dioxide (CO_2) (kar-bun dye-OX-side)
A gas you breathe out. It also makes soda fizzy and raisins dance.

Cilia (SILL-ee-uh)
Tiny hairs inside your nose that sweep out dust, dirt, and invading boogers.

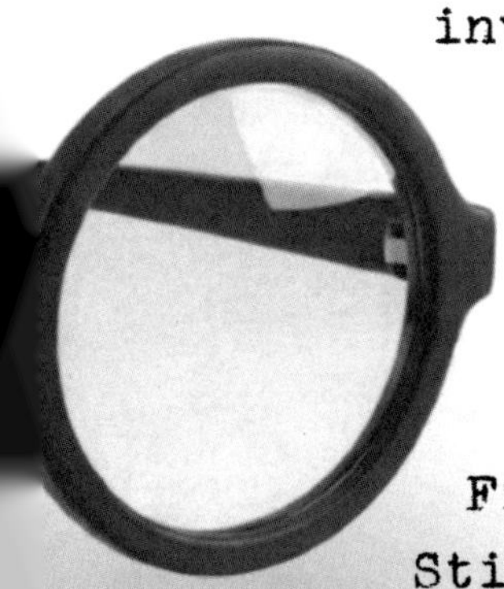

Platelets (PLATE-lets)
Small cell fragments in your blood that help form clots when you get a cut.

Fibrin (FYE-brin)
Sticky protein threads that weave together to make a scab.

Polymer (PAH-luh-mer)
A long chain of connected molecules - the reason slime, snot, and Jell-O stretch.

Viscosity (vis-KOSS-ih-tee)
How thick or sticky a liquid is - honey has high viscosity, water has low.

BONKERS FACTS! VOLUME 2

Coming Soon?

A NOTE TO PARENTS & TEACHERS

Hello responsible grown-ups!

Don't worry! Everything in here is scientifically accurate, educationally appropriate, and only mildly unhinged. Gross facts help demystify the natural world, reduce fear, and build strong science vocabulary.

We promise no actual children will be encouraged to touch ticks, pet hyenas, or test the pH of penguin poop without an adult present.

No ticks, bats, leeches, or imaginary hippos were harmed in the making of this book. In fact, all animals shown in this book were drawn, not handled.

This book supports STEM learning, vocabulary, curiosity, and occasional delighted shrieking. Please supervise all experiments, real or imaginary.

Designed for curious readers ages 6-9.
Approx. Reading Level: Grades 2-4 (Lexile ~600L)

For families and educators, this book works well for:

- Independent reading (Grades 2-4)
- Read-alouds (K-3)
- Classroom science lessons
- Reluctant reader motivation
- STEM enrichment and discussion

This book celebrates diverse young scientists and encourages every child to see themselves as a curious problem-solver. Thank you for supporting all tiny scientists with big questions.

If you have questions, classroom ideas, or student "gross fact submissions," we'd love to hear from you at:
www.bonkersindustries.com

KABOOOOOOOM!